Science Investigations

MAGNETISM:
AN INVESTIGATION

JOHN STRINGER

PowerKiDS
press.

New York

Published in 2008 by The Rosen Publishing Group, Inc.
29 East 21st Street, New York, NY 10010

First Edition

The publishers would like to thank the following for permission to reproduce these photographs:
Corbis: 9 (Richard T. Nowitz), 16 (Peter Beck), 23 (Kevin Fleming), 24 (Steve Raymer), 26 (DK Limited); Ecoscene: 4 (Eric Needham), 29 (Kevin King); Last Resort Picture Library: 19; OSF/Photolibrary: 6 (Hicker/Accent Alaska Rolf), 10 (Plainpicture), 11 (Yoav Levy), 20 (Warren Faidley), 25 (Maruan Bahrour), 27 (GMBH IFA-BILDERTEAM), 28 (Phototake Inc); REX Features: 7 (Garry Gay), 22 (Mykel Nicolaou); Science Photo Library: 4 left (Charles D. Winters), 4 right (Sinclair Stammers), Cover and 8 (Tek Image), 17 (Cordelia Molloy), 21 (Peter Menzel); WTPix: 12.

Editors: Sarah Doughty and Rachel Minay
Series design: Derek Lee
Book design: Malcolm Walker
Illustrator: Peter Bull
Text consultant: Dr. Mike Goldsmith

Library of Congress Cataloging-in-Publication Data

Stringer, John, 1934-
    Magnetism : an investigation / John Stringer. — 1st ed.
        p. cm. — (Science investigations)
    Includes bibliographical references and index.
    ISBN 978-1-4042-4288-3 (library binding)
    1. Magnets—Juvenile literature. 2. Magnetism—Juvenile literature. I. Title.
    QC757.5.S77 2008
    538—dc22
                                        2007032611

Manufactured in China

Contents

What is a magnet?   4

Is the Earth a magnet?   6

What happens inside a magnet?   8

What is a force field?   10

How does its shape affect the poles of a magnet?   12

Are all metals magnetic?   14

Can a magnet work through things?   16

In what other ways do we use magnets?   18

Are there other invisible forces?   20

How are magnetism and electricity connected?   22

What is an electromagnet?   24

Can a magnet make electricity?   26

How do electrical generators work?   28

Glossary   30

Further information   31

Index and Web Sites   32

What is a magnet?

Magnets are mysterious. They may look unexciting, but they have the power to move things without touching them. Magnets draw some metals into their clutches. They may push other magnets away. In some parts of the world, you can find natural magnets. A rock called magnetite, or lodestone, can attract small pieces of metal. The Ancient Greeks knew about lodestones and many such lodestones were found in Magnesia, now the town of Manisa in modern Turkey. It is from here that magnets get their name. Magnetite contains a lot of iron.

Modern magnets are usually made from pure iron. They can also be made from steel, which is an alloy of iron—iron mixed with other metals. Other magnets are made from rare metals, such as cobalt, cadmium, and nickel. But not all metals can be made into magnets.

This magnet is attracting iron. What effect does it have on the copper?

Magnetite, sometimes called lodestone, is naturally magnetic.

INVESTIGATION

What do magnets attract and repel?

MATERIALS

Two magnets and a collection of small objects. Don't use a watch or a credit card!

INSTRUCTIONS

Divide your collection into things a magnet will attract, and things it will not attract. What do you notice about the things that are attracted? Are all metals magnetic? Do the magnets attract one another? Is this always true?

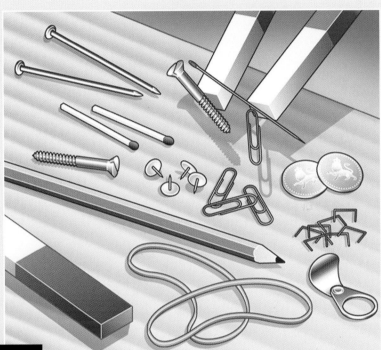

FURTHER INVESTIGATION

How far away will your magnet work? Try putting something small and magnetic at the end of a ruler. Slide your magnet along the edge of the ruler toward it. When does it jump to the magnet? Measure the distance and record your results.

Test the strength of different magnets. Repeat the further investigation, but instead of using the same magnet and different objects, use the same object (for example, a paper clip) and different magnets. The farther away the magnet is when the clip jumps, the stronger the magnet.

ITEM	DISTANCE IT JUMPS
Paper clip	$^1/_2$ in. (12 mm)

Is the Earth a magnet?

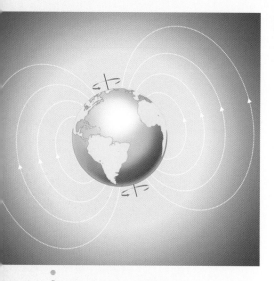

The Earth behaves as if it is a giant magnet. The magnetic poles are not at the true poles of the Earth. These true poles are the North and South Pole.

EVIDENCE

If you hang a bar magnet on a string, it is free to twist and turn. But you will find that it always comes to rest pointing the same way. This is because a magnet is acting as a kind of compass. A compass turns to face north-south, because the Earth itself is a giant magnet.

Around every magnet there is a "field" where the invisible force of magnetism is operating. The Earth has its own magnetic field and magnetic poles. The magnetic north and south poles are where the Earth's magnetic force is strongest. They are in almost the same places as the true north and true south of the Earth (called the *North Pole* and *South Pole*), which are the center of the Earth's spin.

Compasses have a magnetic needle in them that is free to swing around. One pole of this magnet—the "north-seeking" pole—turns to face north. Wherever you are on Earth, you can always find magnetic north on a compass.

The aurora borealis, or northern lights, is caused by the solar wind, a stream of charged particles from the Sun. They are guided by the Earth's magnetic field toward the magnetic poles, where they bombard gases in the upper atmosphere causing them to give off light.

INVESTIGATION

How do you make a compass needle?

MATERIALS

A strong magnet, a sewing needle, a cork or a piece of styrofoam, and a bowl of water.

INSTRUCTIONS

Check that your needle is not a magnet by seeing if it attracts other needles.

Using the magnet, stroke the needle from one end to the other, using one pole of the magnet. Always use the same pole and stroke in the same direction. Try 50 strokes. Find out if you have made your needle into a magnet by testing it.

You can make your own ship's compass. Attach the needle to a piece of cork or styrofoam floating in a bowl of water. It should turn to face north-south.

Magnetic compasses still show sailors the direction at sea. They float on liquid so they can rock with the ship.

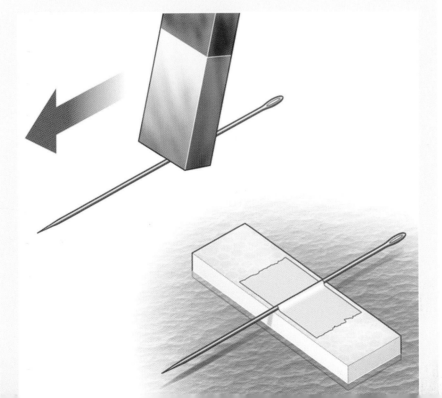

FURTHER INVESTIGATION

Check your floating needle with a magnetic compass. What happens if you bring a magnet close to your floating needle? What will your magnet do to it?

What happens inside a magnet?

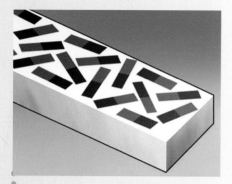

In unmagnetized material, the mini-magnets all point in different directions and cancel each other out.

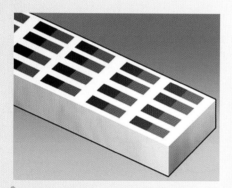

In magnetized material, the mini-magnets line up so they point the same way.

A cluster of filings being attracted by a magnet. The individual filings become magnets themselves.

Iron bars are made up of tiny particles. These particles are grouped together. Each little group is a mini-magnet or domain. Each mini-magnet faces in a different direction, like people in a crowded lift.

In a magnet, the mini-magnets all face one way and are working together. They are like a crowd of people in the elevator who have turned to face the door. And because the mini-magnets all work together, the whole bar is a magnet. The mini-magnets at the ends of the bar face out. The magnetic forces in a bar magnet are concentrated at the ends. At the ends, the magnet's power to push and pull is strongest. We call these ends the magnet's *poles*. Every magnet, whatever its shape, has a north pole and a south pole.

INVESTIGATION

Where do the poles lie?

MATERIALS

Three magnetic sewing needles, tissue paper, and a bowl of water.

INSTRUCTIONS

Make three magnetic sewing needles (see page 7). Float them on a bowl of water by placing each needle gently on a scrap of tissue paper. Push the paper so that it sinks, but the needles should still float.

When you have three needles floating by themselves, steer them together to make a triangle. The triangle will only work one way. What does this tell you about the arrangement of the poles?

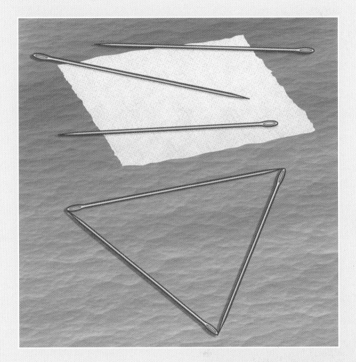

FURTHER INVESTIGATION

Hold a bar magnet horizontally. Try hanging chains of paper clips or safety pins from the ends and from the middle of it. Where are the chains longest?

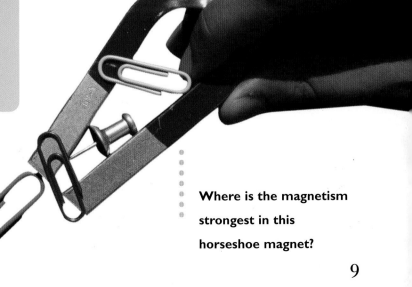

Where is the magnetism strongest in this horseshoe magnet?

9

What is a force field?

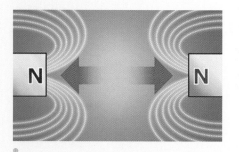

Unlike poles attract. The field between them is strong and the two poles are drawn together.

Like poles repel. When like poles face each other, they push each other apart.

On magnetic levitation railroads, electricity is used to make magnetic power. Magnetism pushes the train upward so it rides above its tracks.

Magnets push and pull things without touching. They can do this because there is a force field around each magnet. In this invisible field, the magnet can attract metal objects, and attract and repel other magnets. The force field can act as an invisible trap. A piece of magnetic metal in the force field of a magnet is drawn to it. There is no escape. The door of a fridge works like this. A long magnet, all along the edge of the fridge, grabs for the metal door as soon as it comes close and pulls it shut.

A force field can also act as an invisible barrier. Bring another magnet into this force field, and it may be attracted and trapped. But it may be repelled, so that you have to push hard to get the magnets to touch each other. Some trains ride on a magnetic force field. Because they don't touch the tracks, they can travel very fast.

If different poles of two magnets are brought together—north to south, or south to north—they attract each other. But if the same poles are brought together—north to north, or south to south—they repel. You know that a metal bar is a magnet when it is repelled by another magnet.

INVESTIGATION

How can you find the force field?

MATERIALS

A bar magnet, a magnetic compass, a sheet of paper, and a pencil.

INSTRUCTIONS

Put the bar magnet in the middle of the paper. Slide the compass close to it. Watch how the compass needle moves. Draw around the compass and pick it up. Now draw the way the needle was facing on the circle. Move the compass and draw the needle again—and again. Notice how the needles are facing.

These directions are the magnet's lines of force. Together, these lines make up the force field.

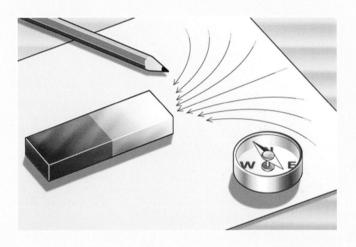

FURTHER INVESTIGATION

Put a magnet under a piece of paper. Sprinkle some iron filings on the paper, and tap the paper gently. Watch as the filings line up to show the force field. Put a wet paper towel on the filings. Wait a day until it has dried out. Blow off the filings and you will have a picture of the force field in rust.

SAFETY: Do not touch the filings. Wash your hands after using them.

· · · · · · · · · · · · · ·

This shows the force field of a bar magnet.

Where is the magnetism concentrated?

11

How does its shape affect the poles of a magnet?

Magnets come in many different shapes. Bar magnets are straight, with a pole at each end—the north and south poles. Some magnets are curved around to make a horseshoe. Both the poles pull together on a metal bar. Some magnets are round like a coin and solid in the center, and others have a hole in the middle. The poles are on the two sides of the magnet. There are also magnetic marbles. These are spherical, but they contain a tiny bar magnet. One side of the marble is the north pole, and the other is the south pole.

Some magnets come in tapes, strips, and strings. There are "rubber" magnets—magnetic strips made from tiny magnets in ribbons of rubber. There are magnetic tapes made from plastic. Look for the brown strip on the back of many identification cards. This strip of magnetic tape identifies bank and club membership cards, so cash machines and locks can recognize them.

Look for the poles on these magnets. How do the magnets get their names?

Portable cassette players like this one store sound as magnetic patterns on tape.

12

INVESTIGATION

Can a magnet "float on air"?

MATERIALS

Three ring magnets and a pencil.

INSTRUCTIONS

Thread one of the magnets onto the pencil. Then thread on the other magnet. What happens? Now turn the top magnet over and thread it back on the pencil. What happens now? What would happen if you put a third magnet on top of the other two? Would it matter which way up it was?

Why do the magnets "float"? Find their poles. Which poles face and repel each other?

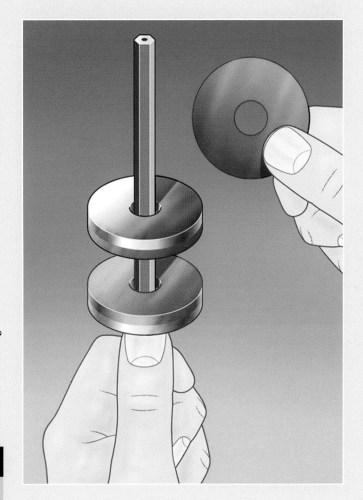

FURTHER INVESTIGATION

Try the investigation again, but this time add weights (for example, disks of card—each the same weight) to the floating magnet and measure the space between the two magnets each time. Draw a graph of distance against weight. What do you think the shape of the graph tells you?

Look at the magnetic strips on cards. Why shouldn't you bring a magnet near them?

Are all metals magnetic?

Everything is magnetic, but the magnetic pull on most objects—including you—is very slight. It would take a very powerful magnet to attract you. Iron and steel, cobalt, cadmium, and nickel are very magnetic. But other metals can only be attracted by hugely powerful magnets. So it is easy to separate steel and aluminum cans. Steel cans are attracted to a magnet, but aluminum cans are not.

Small changes to metals can make them magnetic or nonmagnetic. Steel is an alloy of iron, and it is magnetic. Steel is iron with a tiny amount of carbon and other chemicals in it. But stainless steel—which has a tiny amount of chromium in it, too—is not magnetic. So it can be separated from ordinary steel in a scrapyard, using a magnet.

Huge magnets that are powered by electricity can separate metals in a scrap metal yard.

INVESTIGATION

Which cans are magnetic?

MATERIALS

A magnet and some empty food and soft drink cans.

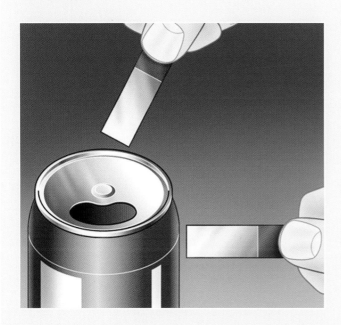

INSTRUCTIONS

Divide a collection of drinks and food cans into steel and aluminum groups, using a magnet. Test the ends as well as the sides— occasionally, the ends of an aluminum can are made of steel. Cans are sometimes called "tins," because the steel can is covered with a thin layer of tin (since the food would cause the steel to rust). But tin metal is not magnetic—the steel inside is what sticks to the magnet.

Now sort the cans according to how magnetic they are.

FURTHER INVESTIGATION

Try sorting soft drinks cans another way— using unopened cans. Put them in a bucket of water. What do you notice? Try the magnet on the cans. What do you notice now?

What metals do you think these objects are made from? Why has each metal been chosen?

Can a magnet work through things?

A magnetic field is invisible. It can work right through some other objects. You can make something magnetic move on a tabletop by attracting it with a magnet underneath the table. You can make it stick to the bottom of the table by putting a magnet on the tabletop—if it is light enough.

The magnetic field works through the table. A magnetic field can work through you, too. Magnetic Resonance Imaging (MRI) uses powerful magnets to build up a picture of the inside of a human body. The magnetic field passes though your body, so that doctors can "see" inside you. A magnet can work through things to make other temporary magnets. If you hang a paper clip from a magnet, it temporarily becomes a tiny magnet, and you can hang another paper clip from it.

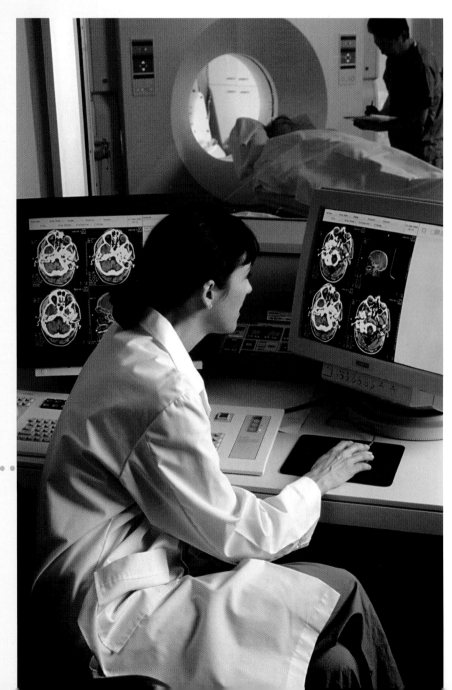

This is an MRI machine. Huge magnetic coils inside the machine surround the patient. MRI scanning uses radio waves and a magnetic field to produce images of the body.

16

INVESTIGATION

Can you do the Indian rope trick?

MATERIALS

A magnet, a paper clip, some thread, and tape.

INSTRUCTIONS

Tie a paper clip to the end of a thread. Tape the other end of the thread to the table. Now lift the paper clip with a strong magnet. Pull the paper clip from the table until the thread is stretched out straight.

Now gently pull the magnet away from the paper clip. A small gap will appear. The stronger your magnet, the bigger the gap.

What can you put through the gap without the paper clip falling down?

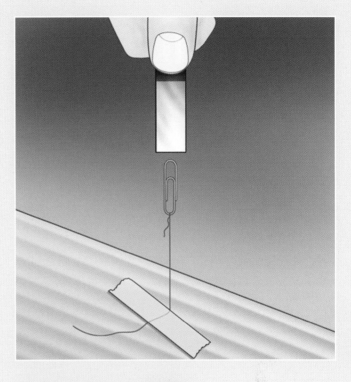

FURTHER INVESTIGATION

Why won't a steel object go through the rope trick gap without the clip falling down?

Can a magnet work through water? Put an object that is attracted by a magnet in water. Put the magnet against the outside of the container. What happens? Try putting the magnet into the water. Does it attract through the water?

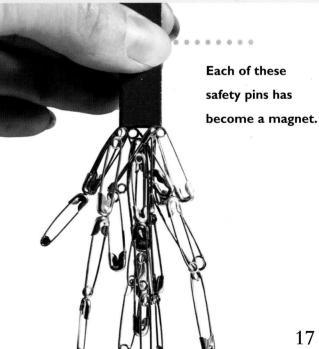

Each of these safety pins has become a magnet.

In what other ways do we use magnets?

There are magnets all around you. There are hundreds in your home. Look for the latches on kitchen cupboards, magnets to hold kitchen knives, and the magnets that keep messages attached to the fridge door. Some household tools such as screwdrivers are magnetic, as well as many toys and games.

Many recording devices, such as cassettes, videos, and computer disks, use magnetic tape where sound is recorded as a magnetic pattern. A minidisc (MD) player records sounds onto a special magnetic layer by heating the disc with a laser. "Smart" cards, which do not need to be swiped to pay for something, work by magnetism. The card sends out a weak magnetic signal picked up by a special aerial. Cards like these will let people on the bus or onto a train platform, and take the money for the ticket from their bank.

You are surrounded by magnets in your home and school.

18

INVESTIGATION

Do a magnet survey

Look for magnets around your home. You should find plenty—especially in the kitchen! Make a list of the ones you find and the jobs they do around the home. Electric motors work by using one magnet to turn another, so everything that contains a motor—from the big one in the washing machine, to the small one in the coffee grinder—also contains magnets.

MAGNET	WHAT THE MAGNET DOES
Door latch	keeps the door shut

FURTHER INVESTIGATION

Ask an adult to help you look for magnets in garages or toolsheds. Combine your investigations with some of your friends' surveys. How many of the same findings do you have?

Minidisc (MD) players record sound as digital code on a magnetizable layer of the disc when it is heated by a laser.

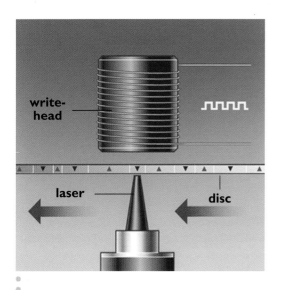

During recording, the laser heats the layer of the disc to be recorded, and a magnetic write-head changes the pattern of magnetization.

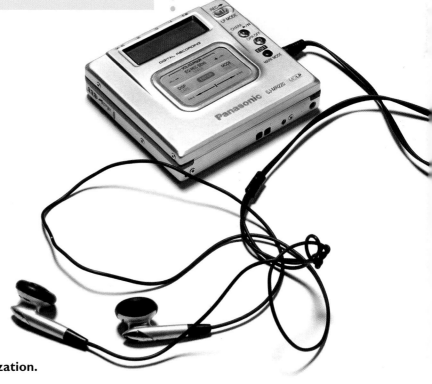

Are there other invisible forces?

Magnetism is not the only invisible force that can attract or repel. A piece of amber can lift hair and attract tiny scraps of tissue paper. When you rub it, it makes electricity (the word for "electricity" comes from the Greek word for amber, *elektron*).

You can do the same thing with your comb, or with a balloon. Rubbing creates static electricity—which means it stays still.

Static electricity attracts light objects, such as dust, scraps of paper, and your hair. But electricity and magnetism are more closely connected than simply both being able to attract (or repel) other objects. Together they make one of the strongest forces in the Universe. It was only after it was understood how they worked together that it became possible to generate electricity as a current.

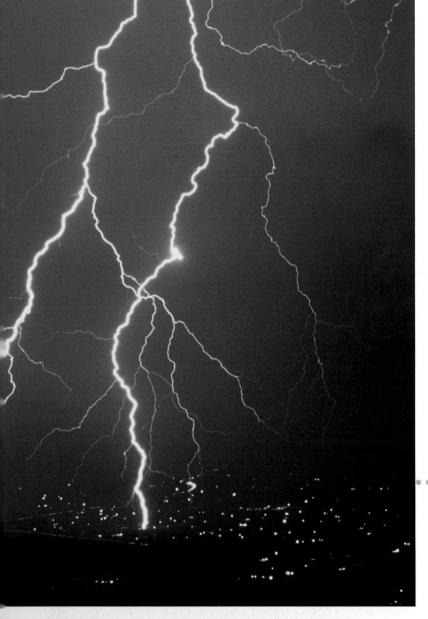

A *thunderstorm* is the natural discharge of static electricity between clouds and the Earth. Long before electricity was created in wires, scientists had learned how to create static electricity for themselves.

INVESTIGATION

How is static electricity like a magnet?

MATERIALS

Two inflated party balloons, some thread, and some woolen material (scarf or sweater).

INSTRUCTIONS

On dry days, you can rub a balloon on your sweater and stick it on the wall, or lift your hair in the air.

Tie a thread on each of two round balloons, and hang one under a table. Rub it about 20 times with the woolen material. Now rub another balloon, hold it by the thread, and bring it close to the first balloon. What happens?

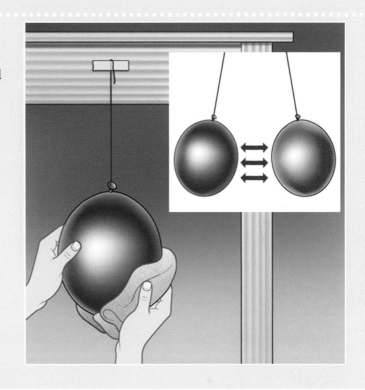

FURTHER INVESTIGATION

Comb your hair, or rub your comb hard with a cloth. Run a faucet so that you get a thin stream of water. Bring the charged comb close to the stream and watch it bend. Why does this happen?

The power of static electricity. A charge passes from the sphere of this machine to this girl's hair. As the hairs become charged, they repel each other and stand on end.

How are magnetism and electricity connected?

Scientists knew of other natural things that could produce electricity in other ways than just by static. The electric eel, for example, can make a powerful electric shock that stuns smaller fish. The electricity comes from special disk-shaped electric organs in the fish.

The first battery was made of a pile of silver and zinc disks, touching both each other and a piece of wet leather. This gave a powerful electric shock. But nobody connected electricity with magnetism for many years. It was only when a scientist brought a magnetic compass close to the wire with an electric current that the connection was made. The compass moved to line up with the wire. The wire had become a magnet.

There is a magnetic field around any wire carrying an electric current, and this field can make magnets move. Instruments that measure the strength of electricity work like this. The more the magnetic needle jumps, the stronger the electricity.

Thousands of power cables run through this busy office, each one generating its own magnetic field. These fields can interfere with the flow of electronic data.

INVESTIGATION

Can you move a compass with an electric current?

MATERIALS

A magnetic compass, a yard of thin insulated wire, and a flashlight battery.

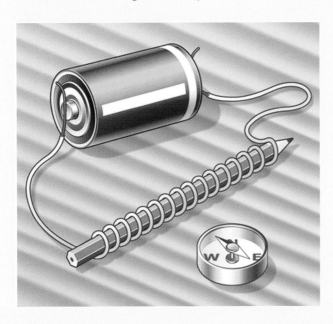

INSTRUCTIONS

Coil the wire around a pencil and touch both ends to the terminals of the battery. Bring a compass close to the loop of wire and its electric current. What happens to the compass?

DO NOT hold the wire to the battery for too long. You are making a "short circuit" and the battery will run down very quickly. The wires might get hot, too!

FURTHER INVESTIGATION

Make a floating compass needle (see page 7). Wait until the needle lines up north-south. Fix a wire across the top of the bowl in the north-south direction. Touch the ends of the wire to the terminals of the battery. What happens to the needle when the electricity is on—and off?

This electric eel makes electricity to stun its prey.

23

What is an electromagnet?

An electromagnet is a magnet you can switch on and off. An electromagnet without electricity is just a metal bar in a coil of wire. But switch it on, and it becomes a powerful magnet that can lift scrap metal, ring doorbells, or surgically pull a steel splinter from your eye.

Electromagnets have a wire coiled around a metal core. When the electric current is switched on, the coiled wire creates a magnetic field. When the electric current is turned off, an iron core will stop being an electromagnet. But a steel core becomes slightly magnetic. Each time you switch on the electromagnet, the steel becomes more and more magnetic.
This is how permanent magnets are made. A metal bar is put inside a strong coil. The electricity is switched on, and the metal bar becomes magnetized. The stronger the current—and the longer the bar is in the coil—the stronger the magnet. You can use a coil like this to "repair" weak magnets, too.

• • • • • • • • • • • •

A metal-detector works as an electromagnet. It has a magnetic field around it. Metal objects disturb the magnetic field, which means objects can be located under the ground.

INVESTIGATION

How can you make an electromagnet?

MATERIALS

A large nail, a yard of thin, insulated wire, and a battery.

INSTRUCTIONS

You can make your own electromagnet. Wind the thin insulated wire around a large nail. Make the winding even—like thread on a spool. Touch the two ends of the wire to the terminals of a battery.

DO NOT hold the wires to the battery for too long. You are making a "short circuit" and the battery will run down very quickly. The wires might get hot, too!

Use your electromagnet to pick up pins and paper clips. What happens when you take the wires away from the battery?

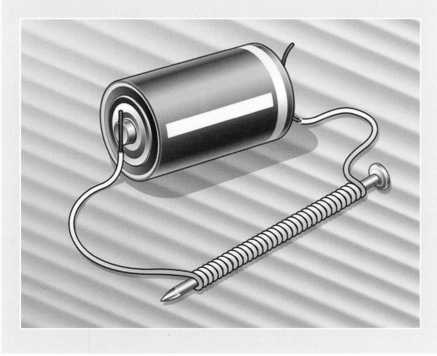

An electric guitar uses electromagnetism. Plucking the strings makes them vibrate in the magnetic field of pickups beneath the strings. This changes the amount of electricity in the wire coils of the pickups, creating electrical signals that run along the lead to boost the sound signal.

FURTHER INVESTIGATION

Use a longer insulated wire for your electromagnet. Increase the number of coils of wire around the core. Find out if the magnet is stronger.

Why do more coils make the electromagnet stronger?

25

Can a magnet make electricity?

Your television, your computer, your washing machine, and many other household items all use electricity that comes to your house through cables. At the other end of those cables is a power station. The power station generates electricity for you to use.

The first scientist to think of generating electricity from magnetism was Michael Faraday. He knew that a current could create a magnetic field, so he wondered if a magnet could create a current. He discovered that if he moved a magnet through a coil of wire, an electric current was produced in the wire.

A simple example of this is the dynamo on a bicycle wheel that powers a light. As you cycle along, you are providing the movement, and the moving magnets in the dynamo generate the electricity to light a bulb. This is called *direct current*, because it always flows in the same direction.

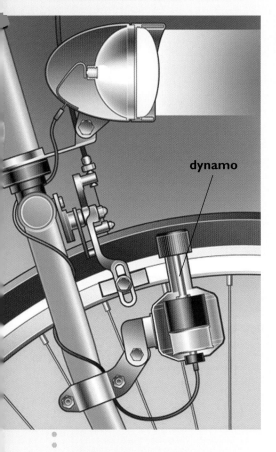

dynamo

A *dynamo* is made up of an electromagnetic coil and a magnet that spins around, inducing electricity to light the bulb.

This early generator or dynamo has two big copper wire coils at the top and bottom. Spinning the magnet (mounted on the rod in the middle) produces electric current in the wire.

26

INVESTIGATION

Can you turn a motor and create electricity?

MATERIALS

A small electric motor, two wires, and a low-voltage flashlight bulb in a bulb holder.

INSTRUCTIONS

Join each wire to the terminals of the electric motor. They will probably be metal tabs on the end of the case. Attach the wires to the bulb holder and screw them in place.

Now hold the drive shaft of the motor between your finger and thumb, and spin it quickly. Watch the bulb. You are moving the magnets in the motor and generating a tiny current. But not very much! The bulb will glow weakly for a moment.

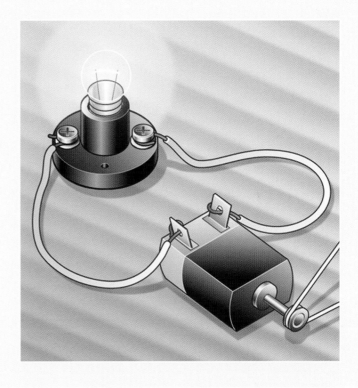

FURTHER INVESTIGATION

Think of other ways of turning the drive shaft in order to make the motor work!

• • • • • • • • • •

Electricity pylons carry the cables from the power station to our homes.

27

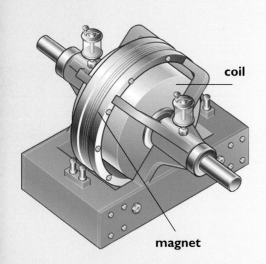

coil

magnet

How do electrical generators work?

Electricity generators contain magnets. When the magnets move, electricity is generated. Most of our electricity generators work by spinning coils of wire between the poles of a magnet. When the wire coil turns in the magnetic field, electricity is produced in the wire.

An alternating current generator reverses the direction of its current every half-turn. The spinning speed of the coil controls the rate that the current reverses.

In a power station, after each turn of the coil, the magnetic field reverses and then returns to its original position. This causes the current to pass first one way and then the other. It is called an *alternating current*. Our mains electricity is supplied by an alternating current generator. Alternating current is safer than *direct current*, which flows in one direction only. Touching any mains electricity can be lethal, but alternating current will break your grip. Mechanical power is needed to turn the coil to create electricity. The pressure of steam (produced by burning fuel), moving water, or the wind can be used. These drive a turbine (a wheel that revolves to create mechanical energy) to turn the generator coils around.

Inside the generators in most power stations are huge magnets that rotate inside a ring of coils.

INVESTIGATION

Can you spin the motor using a turbine?

MATERIALS

A small electric motor, a dish detergent bottle, and scissors.

INSTRUCTIONS

You can spin the motor you investigated using water from the faucet. Cut a turbine from a detergent bottle by cutting across the bottle with scissors about 4 in. (10 cm) from the top. Cut L-shaped slots into the top and bend these out to make paddles. Jam the bottle spout on the motor shaft.

When you put the turbine under the faucet, it should spin the motor and generate electricity by lighting the bulb. This is how a hydroelectric power station works.

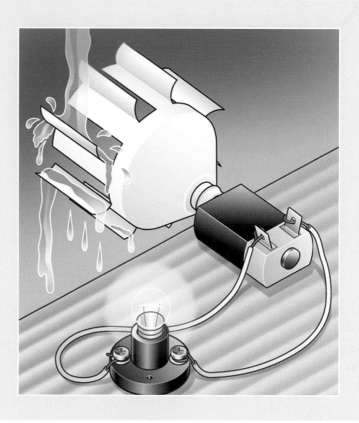

FURTHER INVESTIGATION

Experiment with running the water faster or slower. What effect does this have on what the motor can do?

The power of wind can drive a turbine to produce mechanical energy.

Glossary

Amber
Fossilized tree resin.

Attract
To pull together without touching.

Aurora
Streams of light above the Earth's magnetic poles.

Battery
A container of chemicals that push electric current around a circuit.

Compass
A magnet that is free to move and lines up with the Earth's magnetic field.

Core
A metal rod in the middle of an electromagnet.

Current
The movement of electricity through a wire.

Dynamo
A small electrical generator.

Electromagnet
A magnet that operates when electric current flows through it.

Force field
An area around a magnet where the magnetic force works.

Generator
A machine for producing electricity from movement.

Iron
A strong, heavy metal.

Lodestone
A naturally magnetic rock, made mostly of iron, sometimes called magnetite.

Magnetite
A naturally magnetic rock, made mostly of iron, sometimes called lodestone.

MRI
Magnetic Resonance Imaging: a way of looking inside your body using powerful magnets.

Poles
The ends of a magnet where the magnetism is concentrated.

Repel
To push apart without touching.

Spherical
Ball-shaped.

Stainless steel
Steel with a tiny amount of chromium in it.

Static electricity
Electrical charge which remains in one place.

True north
The position of the North Pole. True south is the position of the South Pole.

Turbine
A wheel that turns when water or steam passes through it, and converts the motion into mechanical energy.

Further information

BOOKS

Awesome Experiments in Electricity and Magnetism
by Michael DiSpezio
(Sterling, 2006)

Electricity and Magnetism (Discovering Science)
by Rebecca Hunter
(Raintree, 2004)

Electricity and Magnetism (Science Files)
by Chris Oxlade
(Hodder Wayland, 2005)

Electricity and Magnetism (Science Files)
by Steve Parker
(Heinemann Library, 2005)

Electricity and Magnetism (Tabletop Scientist)
by Steve Parker
(Heinemann Library, 2005)

Magnetism (Science Answers)
by Chris Cooper
(Heinemann Library, 2003)

Magnetism (Straightforward Science)
by Peter D. Riley
(Franklin Watts Ltd, 2003)

The Way Things Work
by David Macaulay
(Houghton Mifflin, 1998)

CD-ROMS

Eyewitness Encyclopedia of Science
Global Software Publishing

I Love Science!
Global Software Publishing

ANSWERS

page 4 The magnet does not pick up the copper since it is not magnetic.

page 5 All the things that are attracted are made of iron or steel—except stainless steel, which is not magnetic. Only iron or steel—and some rare metals—are very magnetic.

Magnets both attract and repel each other. It depends which poles (ends) are facing each other. See page 10.

page 7 The magnet is stronger than the Earth's magnetism, and so the floating needle is attracted to it and turns around.

page 9 The needles only work one way, because the north and south poles must be lined up with each other.

The magnetism is in the poles so the chains attached to a bar or horseshoe magnet will be longest at the ends.

page 11 The magnetic force is strongest near the magnet's poles, where the filings are closest together.

page 12 The magnets are named according to their shape. The magnets are: bar (top), ring (middle), and horseshoe (bottom).

page 13 The poles on ring magnets are on their flat sides. So turning a magnet over faces the "like poles" toward each other. Like poles repel—the top magnet floats on the repelling forces of the magnets. The pencil holds it in this repelling force.

The shape demonstrates the way magnetic force changes with distance—first it is reduced rapidly, then more slowly.

A magnet will wipe the data from a credit or smart card, making it useless.

page 15 Some sink but others float. This can be because the cans are made from different metals, or it can be to do with the weight of the contents of the can, for instance, sugar can make a can weigh more.

Each metal has been chosen for its strength and attractiveness. Silver and gold are beautiful, but soft, metals. Steel is strong; aluminum is light and doesn't tarnish easily.

page 17 A steel object in the gap is attracted to the magnet and the magnetic force is weakened. The clip falls down.

Magnets can work through water. You can make a magnetic diver from modeling clay and put a paper clip inside him.

page 21 The static charges on the balloons repel each other.

Plastic stores static electricity charges. When the plastic is rubbed, it stores a negative charge. When it is brought close to the running water, it produces a positive charge in the closest part of the flow. The opposite charges are attracted—and the water bends!

page 23 The compass is affected by the wire. While electricity is passing through the wire, it becomes a magnet.

The magnetic needle is floating and free to move. When an electric current passes through the wire, it becomes magnetic, and it deflects the needle. Turn the current off, and the needle returns to facing north-south.

page 25 The pins and paper clips fall off because the nail is no longer magnetized.

More coils make the field lines more concentrated so the magnetism is stronger.

page 29 More mechanical energy will make the motor run faster and more effectively.

Index

aluminum 14, 15
attraction 4, 5, 7, 10, 20
aurora borealis 6

bar magnet 8, 9, 11, 12
battery 22

compass 6, 7, 11, 22, 23
copper 4

direct current 26, 28
domain, see mini-magnet
dynamo 26

Earth, magnetism of the 6–7
electric eel 22, 23
electric guitar 25
electricity 10, 20, 22–23, 24, 26–27, 28
    see also mains electricity, static electricity
electricity generators 26, 28–29
electricity pylons 27
electromagnets 24–25

Faraday, Michael 26
force field, see magnetic field

horseshoe magnet 12

iron 4, 8, 14, 24

lodestone 4

magnet, see bar magnet, electromagnets,
    horseshoe magnet, magnetic marble,
    magnetism, ring magnet
magnetic field 6, 10–11, 16, 22, 24, 25, 26, 28
magnetic levitation 10
magnetic marble 12
magnetic strip 12, 13

magnetism
    and electricity see electricity, electromagnet
    and metals 4, 5, 10, 14–15
    introduction 4–5
    through materials 16–17
    uses of 10, 12, 14, 15, 18–19, 24, 26
    what happens inside a magnet 8–9
    see also Earth, magnetism of the
magnetite see lodestone
mains electricity 28
metal-detector 24
minidisc player 18, 19
mini-magnet 8
MRI 16

northern lights, see aurora borealis

poles 6, 8, 9, 10, 12, 13

repulsion 5, 10, 20
ring magnets 12, 13

static electricity 20–21, 22
steel 4, 14, 15, 17, 24

wind farm 29

Web Sites

Due to the changing nature of Internet links, PowerKids Press has developed an online list of Web sites related to the subject of this book. This site is regularly updated. Please use this link to access this list:
www.powerkidslinks.com/sci/magnet